Gorilla

STRUGGLE FOR SURVIVAL IN THE VIRUNGAS

Mount Mikeno, "The Naked One," 4,432 meters high, was formerly an active volcano, whose eroded slopes form part of the ancestral home of the mountain gorilla.

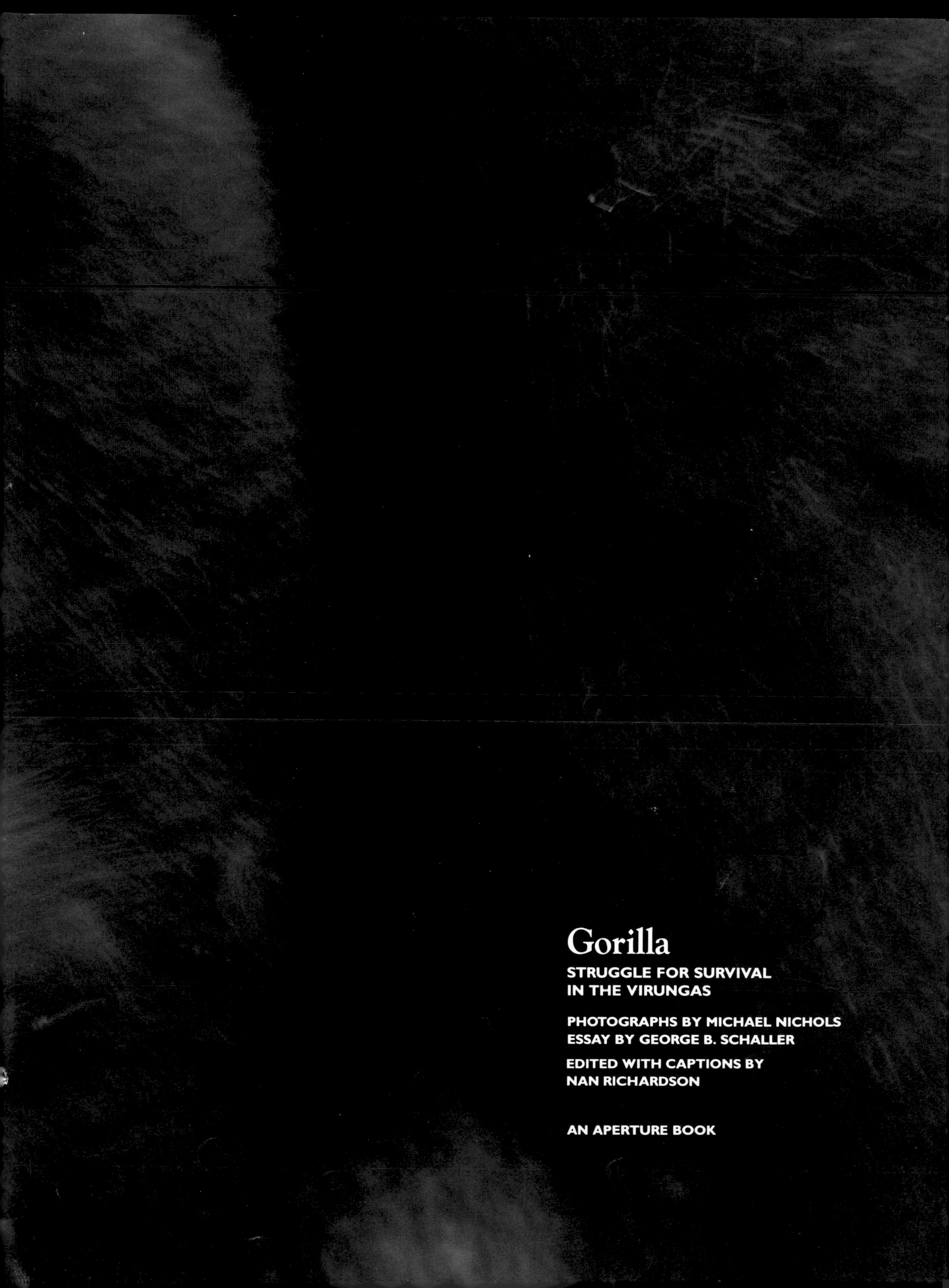

Gorilla

STRUGGLE FOR SURVIVAL IN THE VIRUNGAS

PHOTOGRAPHS BY MICHAEL NICHOLS
ESSAY BY GEORGE B. SCHALLER
EDITED WITH CAPTIONS BY
NAN RICHARDSON

AN APERTURE BOOK

In Africa, the preserve of the Virunga Volcanoes forms the last stronghold for the few hundred remaining mountain gorillas in existence. Straddling the common border of Rwanda, Zaire and Uganda, the rainforest and rich biosphere in which they live shares a common destiny with the lives of its inhabitants. This is the story of a hard-won balance in conservation: a people, a land and a magnificent primate, one of man's closest relatives. It is a story with lessons—and warnings—for our future on this earth.

 Distributed in the United States by Farrar, Straus and Giroux. Library of Congress Catalog Number: 88-70941.
ISBN: 0-89381-310-9 (Cloth)
ISBN: 0-89381-349-4 (Paper)

Composition by Trufont Typographers, Inc., Hicksville, New York. Printed and bound in Hong Kong by South Sea International Press Ltd. Book and jacket design by Betty Binns Graphics/David Skolkin. Published in France by Editions Nathan, Paris.

Aperture Foundation publishes a periodical, books, and portfolios of fine photography to communicate with serious photographers everywhere. A complete catalog is available upon request. Address: 20 East 23rd Street, New York, New York 10010.

"This pot-bellied monster, a 360-pound gorilla nearly six feet tall, once shuffled through the highland bush of central Africa. Hunter Herbert Bradley, right, shot the giant in 1921. Naturalist Carl Akeley, next to Mrs. Bradley, cups the gaping jaw. "Mrs. Bradley, Mr. [Carl] Akeley and Mr. H.E. Bradley, Belgian Congo, Africa, 1921," reads the legend. Courtesy the American Museum of Natural History, New York.

Clambering along the steep slopes of Mount Mikeno, I pushed through the undergrowth of thistles, sourdock, and stinging nettles until I came upon a swath of freshly trampled vegetation. The gorilla trail was fresh. The musky, somewhat sweet odor of the animals was still in the air. I followed the trail cautiously until the snap and crunch of feeding gorillas was just ahead. Quietly I pulled myself into a *Hagenia* tree and huddled there trying to become part of the trunk and cushions of moss that padded the branch on which I sat. From my vantage point, I could look over the undergrowth and observe the gorillas as they peeled the succulent stems of wild celery and tore apart branches of the *Vernonia* shrub to gnaw out the tender pith. I watched the apes undetected for over an hour.

But a young male suddenly spotted me and, rearing up on his hind legs, slapped his chest. Mr. Crest, my name for the leader of this group of twenty-one gorillas, jerked to attention. He was a silverback, his gray saddle proclaiming his maturity; browridges protruded over his eyes like a cornice and the crest on his head was huge, resembling a hairy miter. He roared and the forest vibrated from the explosive sound. His family collected around him, but, instead of fleeing, all advanced purposefully to within thirty feet of me. A plump female with a three-month-old infant clutched to her chest angled closer. Peering at me out of the corner of her eye, she gave the low-hanging branch on which I sat a sharp tug, then glanced at me to gauge my response—and then she climbed up on the branch with me! We sat there a trifle uneasy, like a pair of strangers at opposite ends of a park bench. Curiosity satisfied, she descended. Her place was briefly taken by an adolescent and then by another female. Interest in me waned after that, the gorillas continuing their routine as if I did not exist.

Nearly three decades have passed since I shared that branch with wild mountain gorillas, yet the exhilaration I felt at being accepted by the apes is with me still.

Gorillas exist in West Africa and east-central Africa, the two areas separated by about 600 miles of tropical rainforest. When in 1959 I began a study of the mountain gorillas in the Virunga Volcanoes, which straddle the international boundaries of Zaire (then the Belgian Congo), Rwanda, and Uganda, little research had been done on the habits of gorillas. In fact, they had remained unknown to the Western world until discovered in 1847 by Thomas Savage, a missionary in West Africa. He wrote that "they are exceedingly ferocious, and always offensive in their habits." In 1856, the American explorer Paul du Chaillu was the first Westerner to shoot a gorilla, a "hellish dream creature—a being of that hideous order, half-man half-beast, which we find pictured by old artists in some representations of the infernal regions." Even scientists accepted such skewed perceptions, and in 1859, for example, the British anatomist Richard Owen wrote:

"Negroes, when stealing through shades of the tropical forest, become sometimes aware of the proximity of one of these frightfully formidable apes by the sudden disappearance of one of their companions, who is hoisted up into the tree, uttering, perhaps, a short choking cry. In a few minutes he falls to the ground a strangled corpse."

Humankind projects onto animals its desires and fears and in the end observes mainly the fiction it has created. In the black countenance and tremendous strength of the gorilla it sees less an animal than a myth, a mysterious and monstrous image of itself.

In 1902, Oscar von Beringe, a German officer traveling through Rwanda, came to the eight Virunga Volcanoes. These stretch like a gigantic dam across the floor of the Albertine rift, a huge ditch in the earth's crust flanked by rugged escarpments. The two westernmost volcanoes are still active, whereas the fires in the eastern six are temporarily damped. There on the slopes of Mount Sabinyo—the "Teeth of the Old One," as named locally for its five summits—von Beringe spotted "a group of black, large apes which attempted to climb to the highest peak of the volcano. Of these apes we managed to shoot two. . . ." These gorillas had longer hair, a longer palate, and other minor morphological differences from those in West Africa, and consequently two races or subspecies were recognized, the lowland gorilla, pedantically given the scientific name *Gorilla gorilla gorilla*, and the mountain gorilla, *Gorilla gorilla beringei*, named in honor of its discoverer.

When John Emlen, professor of zoology at the University of Wisconsin, and I arrived in the realm of the mountain gorilla, our first task was to determine the distribution of the animal. We traveled widely from the cool highlands of the escarpments bordering the rift valley to the enervating heat of Zaire's lowland rainforest, where beneath cloudlike billows of foliage no breeze stirs and rain squalls leave the earth sodden and steaming. It soon became clear that most so-called mountain gorillas lived not only in the mountains, as in the Virunga Volcanoes where they roam between 7,800 and 11,000 feet, but also at low elevations. They were not scattered at random in the seemingly endless forest but concentrated in numerous more or less isolated pockets, often in or near the abandoned fields of villagers. In the dusky interior of undisturbed rainforest, between the spectral trunks of huge trees, vegetation near ground level is scant; there is little gorilla food. But in abandoned fields, with their riot of vines, saplings, and herbs, gorillas find the leaves, roots, stems, fruits, and other items they relish. Paradoxically, the shifting cultivation practiced by local people provides gorillas with ideal habitat, and it no doubt favored an increase in their numbers after agriculturalists penetrated the forest. Occasional enmity erupts between these people and the gorillas, usually after the latter have raided a banana grove, where they leave the fruit but tear apart the stem to devour its pith. Villagers may then band together to spear or shoot the animals, spurred on in this retaliation by the promise of feasting on the meat of their victims. But gorillas may in turn defend themselves, sometimes wounding or even killing an attacker. However, there are still relatively few people in that forest and commercial exploitation for timber is not yet extensive: so far the gorillas have remained secure.

Since the time of our 1959 survey, taxonomists have split what was called the mountain gorilla into two races. Now the mountain gorilla, *beringei*, is said to occur only in the Virunga Volcanoes and the nearby Impenetrable Forest of Uganda. All others in east-central Africa are of the race *graueri*, Grauer's gorilla. This change represents more than a taxonomic game with bones and calipers. If only one race is recognized then there is no urgent conservation problem, there still being several thousand animals in Zaire's lowlands. But if there are two races—as is now generally accepted—the mountain gorilla is in serious trouble with only a few hundred left in two forest patches.

First on the spot after the discovery of the mountain gorilla was an army of museum collectors, among them Carl Akeley who in 1921 shot five for the American Museum of Natural History. But so impressed was he with the beauty of the Virunga Volcanoes and the gorillas there that he urged the Belgian government to set aside a sanctuary. Albert National Park was established in 1925. With the independence of the colonies, Albert Park became Parc National des Virunga on the Zaire side of the border and Parc National des Volcans on the Rwanda side. In addition, Uganda established the small Kigezi Gorilla Sanctuary at the eastern corner of the range. Akeley was one of the first to raise his voice on behalf of the gorilla, emphasizing that it was "normally a perfectly amiable and decent creature." In 1926, he returned not to shoot but to study the animal. After reaching a small meadow named Kabara, in the 10,000-foot saddle between Mount Mikeno and Karasimbi, he died. His wife buried him at the edge of the Kabara meadow with a view of Karasimbi's lovely volcanic cone, its summit sometimes capped with snow. Thirty-three years later my wife Kay and I lived near his grave in a park rangers' hut doing the research that was denied him.

Gorillas are far more placid and reserved creatures than even Akeley suspected. I soon learned that the best way to observe them was to settle myself quietly but conspicuously nearby so that they could inspect me and, I hoped, consider me just another innocuous creature in their environment. Their acceptance of my presence was remarkably rapid. And I learned to understand them. Their soft brown eyes more than anything else informed me of what the animals were thinking, as, like silent mirrors of the mind, they reflected unease, curiosity, boldness, and sometimes annoyance. In the beginning, my eyes, too, must have conveyed unease.

The gorilla's strength and potential to do harm demands respect. A silverbacked male has large canines and he may weigh 400 pounds; just seeing one sprawled at rest is to know the power of nature. But the apes evoked a stronger emotion, as I tried to express in my book *The Year of the Gorilla:*

He lay on the slope, propped on his huge shaggy arms, and the muscles of his broad shoulders and silver back rippled. He gave an impression of dignity and restrained power, of absolute certainty in his majestic appearance. I felt a desire to communicate with him, to let him know by some small gesture that I intended no harm, that I wished only to be near him. Never before had I had this feeling on meeting an animal. As we watched each other across the valley, I wondered if he recognized the kinship that bound us.

For over a quarter of a century it has been known that gorillas are not belligerent beasts, yet, as if this is somehow difficult to believe, almost every article about the apes stresses the fact again. And as I write this, I find myself almost inadvertently doing so too. Admittedly, it gave me at first a vicarious thrill to be near gorillas; I exulted in their acceptance as if I had overcome an enormous cultural barrier, for, of course, it is not so much what we see but what it suggests that arouses us. As the days and weeks passed and I recognized many animals individually by their distinctive faces, especially their noses, and traveled leisurely along their trails, perceiving the world at their speed not mine, the study moved into another dimension. When my notebook began to refer to gorillas by name, the animals ceased to be anonymous members of a species and became individuals, each with foibles, sensitivities, problems, family ties, traditions, and past experiences. Yet I knew that in spite of my emotional involvement and our kinship, I would remain an outsider, vainly trying to understand another culture. Gorillas are neither our brothers nor inferior beings to exalt or denigrate, but creatures with a perfection uniquely their own, adapted to life in the forest. My task was to interpret that life, using fact, empathy, and intuition. I had to observe them without moral judgment, without finding in them what humankind seeks, whether it be a vision of the peaceable kingdom that has eluded us or the search for a demonic spirit to haunt our dreams.

As if to atone for decades of slandering the gorilla's character, the animal is now often pictured as a gentle giant. However, gorillas are not always amiable. The fact that a silverback is twice the size of a female suggests in itself that competition for mates and defense of family are important in his life. Being both leader and protector of his group, a silverback often has to confront his principal enemy—man. But even when harassed he prefers bluff to outright attack, though intent is not always easy to judge. Once, for example, as I sauntered through the forest, my mind adrift elsewhere, the undergrowth swayed fifty feet ahead and several gorillas disturbed in their noon siesta raced silently away. The silverback, however, stepped behind a bower of vines and waited hidden in ambush. I stood motionless. Seconds passed and then minutes, each of us waiting for the other to make the first move. At last, without warning, the silverback rose to his full height, flung his gigantic arms skyward, and gave one shattering roar. Wheeling away, he left me intimidated and uneasily looking at the now empty forest. Had I advanced, he might have charged in typical manner, crashing toward me with terrible speed, emitting screaming roars, to deliver a massive blow and perhaps a bite. In reading accounts of attacks on hunters,

"And now he reminded me of nothing but some hellish dream creature—a being of that hideous order, half-man, half-beast, which we find pictured by old artists in some representations of the infernal regions. He advanced a few steps—then stopped to utter that hideous roar again—advanced again, and finally stopped when at a distance of about six yards from us. And here, just as he began another of his roars, beating his chest in rage, we fired and killed him."

"I traveled always on foot and unaccompanied by other white men—about 8,000 miles. I shot and stuffed and brought home over 2,000 birds of which more than 60 were new species and I killed upwards of 1,000 quadrapeds of which 200 were stuffed and brought home, along with more than 80 skeletons."

PAUL DU CHAILLU

HUNTER KILLED BY A GORILLA.

"Hunter killed by a gorilla" reads the caption from an 1861 woodcut and excerpts (above) from *Explorations and Adventures in Equatorial Africa*, by Paul Du Chaillu, copied from a first edition given to Dian Fossey by David Attenborough. Paul Du Chaillu, the American explorer, arrived in Africa in 1856 and became the first white man to shoot a gorilla. Despite his colorful and highly exaggerated descriptions, which fed the contemporary popular imagination, his account remained one of the most accurate for a hundred years.

photographers, or others, I do so now with the quiet satisfaction of knowing that through luck and politeness I never violated gorilla etiquette to such an extent that I deserved a serious charge.

Among themselves, gorillas settle most disagreements through intimidation rather than battle, often beating their chests to signal tension. I had known that a gorilla hits "his chest in rage," as Du Chaillu phrased it. But I was unprepared for the beauty and complexity of the display. All gorillas, even a tiny youngster barely able to rear on its hind legs, will beat a rapid tattoo on the chest with cupped palms, but only the silverback gives a virtuoso performance. One day I came upon The Climber, a silverback so named because of his unusual predilection for clambering into trees, an exercise bulky males generally forgo. He sat staring at the ground as if in deep thought, his group of eighteen females and youngsters clustered around him. Only twenty feet away was another silverback, leader of a group whose members were scattered in the nearby undergrowth, and he was excited. He emitted soft hoots, faster and faster, then abruptly stopped to pluck a leaf and place it gently between his lips, a puzzling act of irrelevance. But then he exploded in a crescendo of action as he reared up, hurled leaves, pounded a resounding percussion on his chest, and ran sideways, slapping at and breaking anything in his path, to end the display with a vigorous thump on the ground with his hand. John Emlen likened the display to a symphony, beginning with slow hoots, rich in restraint, to the crashing finale. Having shown off, the silverback looked around as if expecting applause for his magnificent performance.

Instead there was silence. The Climber rose and walked to the other male and the two faced a foot apart, staring into each other's eyes. These giants were not settling a difference of opinion by brawling but by playing on each other's nerves. The Climber broke first, returning to his seat. Once more he tried to intimidate his opponent, first by throwing a handful of weeds into the air, then rushing to the other animal to face him browridge to browridge. Once more the bluff failed. The Climber left the area, closely followed by his family. The next morning, trails showed that the two groups had met again and that the two silverbacks had tussled: tufts of hair pulled out by the roots littered the trampled undergrowth. Some group meetings end even less amicably, fights between silverbacks resulting in serious injury.

There were ten gorilla groups in my study area around the Kabara meadow, some using it often, others at long intervals. The smallest group contained eight individuals and the largest twenty-seven. Each was led by a silverback at least twelve years old. Groups also contained a variable number of blackbacks—young males aged eight to twelve whose backs had not yet turned gray—as well as females and their offspring. A few groups had more than one silverback, as many as four, yet one was the undisputed leader, a benign dictator who by his actions determined where to go, how far to travel, when to bed down for the night, and other details in the daily routine. Gorillas lay no claim to a plot of land of their own, the trails of several groups crisscrossing haunts where their favored food grows most profusely. Home ranges of groups encompass from about two to fifteen square miles. The animals have strong attachments to members of their own group, individuals often staying together for years, yet there are shifts in membership. One of my groups was led by Big Daddy, a huge, swaybellied male; also in the group were three other silverbacks, D.J., The Outsider, and Splitnose, and each of them left to lead a solitary life during the year I knew the group. Once a new female with an infant appeared in The Climber's group. I wondered where she had come from, for females, unlike males, do not wander around alone. My study was too brief to unravel such intricacies in the gorilla's social life. However, subsequent research by many investigators, among them Dian Fossey, Alexander Harcourt, Amy Vedder, Alan Goodall and Kelly Stewart, helped me understand why that day The Climber was so uneasy about the persistent presence of the other silverback: he was afraid of losing the allegiance of his females.

Females may emigrate voluntarily from one group to another, usually when two groups meet, behavior quite unlike that of most primates in which males transfer. A female may also join a lone male, providing him with the opportunity to start a family. No wonder I sometimes found a loner hanging around a group, trailing it and sleeping near it, hoping no doubt that a female would find him attractive and powerful enough to trust him with her life. But a male may have a long, solitary wait: most silverbacks spend at least three years alone before acquiring permanent female companionship. Most females transfer at least once, usually before they mate for the first time at about eight years of age.

Since a silverback remains group leader for so many years—his life span may be forty years—he most likely is the father of any maturing female; emigration by females thus prevents incest. Having come together under a new leader, most females in a group are unrelated, and this has a subtle consequence. I had noted that females are highly tolerant of each other, spending all day feeding and resting with seldom a squabble. Yet they rarely made friendly overtures such as grooming each other, behavior so conspicuous in various monkey species. They seemed alone together. Intimate gestures were reserved mostly for kin, especially for an infant.

Early every morning, I went in search of one gorilla group or another, following buffalo trails that wound through undergrowth wet with dew. Sometimes the weather was fine, wind-tattered clouds colliding with Mount Mikeno and white-necked ravens playing airy games above the tree tops. On such days gorillas may stay in bed late, and I came upon them still in their nests long after sunup, waking from a night's sleep. They yawned, stretched, reclined again, but finally they ambled about stuffing their paunches. After feeding, toward mid-morning, they would begin a rest that lasted for as long as three hours. It was my favorite time to be with them. They were utterly relaxed, some sprawled on the ground, others leaning against tree trunks, females and youngsters often crowding the silverback, enjoying his company; on one occasion a female used the broad back of one as a cushion while she slept. All was quiet except for soft grumbles and grunts of content. Sometimes a blackbacked male, restless and reckless, came to inspect me. I particularly remember Junior strutting past me at ten to fifteen feet, his indrawn and compressed lips conveying tension. Then with sly mischievousness he swatted the ground with a wanton thump and cantered off, looking back at me over his shoulder.

A newborn is tiny, weighing only four to five pounds, and its mother carries it at all times carefully clutched to her chest. But when at three to four months of age it can shakily crawl, she may permit it to leave her arms to explore around her, and when she moves she may carry it piggyback, the young clutching her long hair. After that a youngster develops rapidly and seeks out others in play. To watch them frolic during the noonday rest in the security of the group, playing king-of-the-mountain and follow-the-leader, wrestling, and swinging by themselves on vines, provided some of my happiest moments of gorilla watching. Even sedate adults were sometimes coaxed into participating, as described in an incident from my field notes.*

A 6½ month-old infant, Max, leaves his mother and climbs up the body of a reclining female, who has no infant. Max advances onto her head, climbs down, rips off a leaf, and attempts another ascent of the female, who has now rolled onto her back. Finally Max reaches the summit of her abdomen and lies down. The female covers the infant with her hand, holding him down. Then Max struggles and squirms and tries to free himself. One hand emerges and then the other; his mouth is partially open and the corners pulled back in a smile. The female then toys with Max, touching him here and there, and in turn he attempts to catch her elusive hand. . . . Suddenly he sits and with arms thrown over his head, dives backwards into the weeds.

After resting, gorillas fed again and toward evening prepared for bed. The behavior of Big Daddy's group was typical. One day at dusk the gorillas snacked in desultory fashion, looking subdued. Fog drifted around Big Daddy as he sat motionless by a shrub. He reached out and bent a branch toward himself and pushed it under a foot. Slowly rotating, he pulled in all vegetation within reach to construct a crude nest. Then he reclined on his belly, arms and legs tucked in, presenting his back to the rain that had begun to fall. When he bedded down so did the others, youngsters crawling in with their mothers, as they would do until reaching adolescence at about three and a half years. Within minutes, at half-past five, all movement had ceased, and the dark forms of the gorillas fused with the night. The turf at the base of the tree from which I had watched them was dry, protected from rain by the slant of the trunk. There I spread my sleeping bag and lying on my back watched the beardlike lichens swaying gently on the branches above.

When Zaire became an independent country in June 1960, the Belgian park staff fled. Poachers now scoured the forest for duiker and buffalo, and Tutsi tribesmen from Rwanda penetrated into Zaire with large cattle herds. Eating vegetation to a stubble, churning soil with hooves, splattering everything with dung, cattle soon make a habitat unsuitable for gorillas. The new administration acted courageously, confiscating cattle and arresting poachers. However, the Rwandan part of the park, which would remain under Belgian control until that colony achieved independence in 1962, was virtually unprotected. A weak Belgian administration had in 1958 relinquished about twenty-seven square miles of gorilla forest to agriculturalists, and pressure for more land in this densely populated country might soon doom the remaining forest there. With the region in political turmoil, Kay and I could not continue the study. On my final visit, I saw The Climber and his family. I wished the animals luck and a peaceful life. As

*From George Schaller, *The Mountain Gorilla* (Chicago: University of Chicago Press, 1963).

I left them, reluctantly exchanging a year of happiness for treasured memories, I did so with a feeling that I would never return.

Yet three years later I was back on a week's visit to guide a photographer for *Life* magazine. I returned with trepidation, fearful of finding the forest destroyed, but there were no cattle, and Kabara was serene. I found The Climber and his family. He was still the leader, though his shoulders were now gray with age and he seemed less dynamic. Shorthair in this group had evolved from a shrill blackback into a deep-voiced silverback. I encountered Mr. Crest and his family too, but something drastic had happened to change the size of his group from twenty-one to ten. All four groups we contacted were nervous and fled for a mile or more immediately after they met us, climbing high into the ravines of Mount Mikeno. They had obviously been harassed, their mountain peace shattered. Their idyll and mine was over.

For a quarter of a century I have not revisited the gorillas but they live within me to appear at unexpected moments. In the tattoo of a woodpecker's hammering I hear the pok-pok-pok of a chestbeat, and in a forest's shadow I see a gorilla in dark green foliage.

Kabara had an American visitor within a month of my departure in 1963, Dian Fossey on her first pilgrimage to the gorillas. She left "with never a doubt that I would, somehow, return to learn more about the gorillas of the misted mountains." And in 1967 she did return to Kabara, though within a few months a rebellion in Zaire forced her to move to Rwanda where on the slopes of Mount Visoke she established the Karisoke Research Center, funded largely by the National Geographic Society. She was immensely determined. "Much can be gained by crawling, rather than walking, along gorilla trails," she explained in her book *Gorillas in the Mist*, for the apes can be tracked that way by the body odor that lingers just above ground. She devoted herself to the gorillas with a sense of mission, perhaps almost mania, studying them and protecting them. She brought in many scientific coworkers, notably the British zoologist Alexander Harcourt, and together they provided unequaled insight into the ecology and behavior of gorillas.

To study gorillas, their groups were habituated to a much greater degree than were my study animals. Peanuts, a blackback, achieved immortality by being the first wild gorilla to reach out his hand in friendship to a human. As Dian wrote: "Since he appeared totally relaxed, I lay back in the foliage, slowly extended my hand, palm upward, then rested it on the leaves. After looking at my hand, Peanuts stood up and extended his hand to touch his fingers against my own for a brief instant." For a fleeting moment they bridged the gap between ape and human, an experience I envy. Later she and others became almost family members of various groups as they sat among the gorillas to groom and play with them. I wonder, though, if such intimacy is fair to gorillas. Contact could easily transmit an illness from humans to the apes, one to which they lack resistance.

"Song and dance in the village of King Kong, Africa, circa 1900." In this early photograph, women celebrate the death of the beast (a lowland gorilla in this case). From *Südafrika Schwarzafrika 1890–1925: Eine Historische Foto-Reportage*, by Eric Baschet. (Kehl-Rhein: Swan Verlag, 1925) Photograph courtesy Sygma, Paris.

As the years passed, Dian Fossey became more and more involved in what she called "active conservation"; that is, she pursued poachers, burned poacher camps, herded illegal cattle out of the park, and in general became fiercely protective of her self-appointed charges. Her priority was correct: when the existence of a rare creature is threatened, a conservation effort becomes primary, science secondary. And the gorillas now needed all the help they could get.

In 1968, 38.5 square miles, or forty percent of the total park area in Rwanda, was destroyed in a European-sponsored scheme to grow pyrethrum, from which an insecticide is made. It was a time when the dangers of DDT were finally admitted and alternatives sought. But by the time the forests had been cut and a factory built, the market for pyrethrum had collapsed, a synthetic substitute having become available. Now mostly potatoes grow where gorillas once roamed. Illegal cattle grazing was widespread in the Rwanda as well as the Zaire and Uganda parts of the park. However, poaching was the most serious immediate threat to gorillas. In the wake of the 1967 civil war in Zaire, hunters in large numbers entered the park to spear, shoot, and snare wildlife. Snaring is unselective, killing and maiming indiscriminately. One report noted that of eleven gorillas in one group "two have only one hand and a third a deformed hand—the result of early encounters with snares." When a snared gorilla in a panic tears itself free, the wire may cut deeply into its flesh, causing gangrene with later loss of life or limb. As if this were not enough of a problem, the Cologne Zoo convinced the Rwandan authorities to provide it with a mountain gorilla infant. Poachers were hired to capture one, and Dian Fossey later "learned that ten members of the gorilla group were killed in the capture." A second infant was just as senselessly obtained from another group. (Both infants died in 1978 at the zoo.)

In 1960, I estimated that about 450 gorillas inhabited the Virunga Volcanoes. A detailed census conducted between 1971 and 1973 by Dian Fossey, Alexander Harcourt, and their associates revealed a drastic decline to about 275. Gorillas reproduce slowly. If her infant survives, a female will not have another for about four years. Some forty to fifty percent of the young die before reaching adulthood, most in the first year of life, from respiratory illnesses, accidents, and other causes. When a silverback is killed by poachers—as he often is in trying to protect his family—the disruption to the group caused by his death can have serious consequences. When a new silverback takes over a group he may kill all infants. Such infanticide is an important reproductive strategy among various social mammals, including langur monkeys, lions, and prairie dogs to name a few. By eliminating young sired by a predecessor, a new male stimulates females to come into estrus soon and mate with him, conceiving his progeny, perpetuating his genes.

I had seen no evidence of infanticide in the stable, peaceful groups I observed. But with the heavy poaching and consequently the frequent changes in group leadership, a quarter of all infants born were bitten to death. Groups in the 1970s had noticeably fewer youngsters with them than during earlier years. Average group size around Kabara dropped from about seventeen to nine.

In 1974, the Rwanda government strengthened the guard force in the Virunga Volcanoes and finally began to remove illegal cattle, fining owners the equivalent of ten dollars per head, a policy that soon eliminated that problem, though not on the Zaire side where park officials were still apathetic. But now a new and alarming threat menaced the gorillas. Tourists and white Rwandan residents bought gorilla hands and heads from poachers as souvenirs. That anyone could aspire to such a grisly trophy conveys something about humankind. Dian Fossey saw Digit, a male she had known "as a playful little ball of black fluff ten years earlier," with his head and hands hacked off, and she noted, "I came to live within an insulated part of myself." She seemed to have lost everything, even despair.

Silently, as if entombed by fog, the Virunga gorillas might have retreated into oblivion had not Dian Fossey drawn international attention to their renewed plight. Now a major effort is launched to help them, one that actually had its beginnings in another part of the gorillas' range.

High on the escarpment west of Lake Kivu, about sixty miles southwest of the Virunga Volcanoes, is the Parc National du Kahuzi-Biega, named for the two highest peaks in the area. In 1966, Adrien Deschryver, a Belgian administrator in the forest service who had remained after independence, noted that agriculturalists were rapidly converting forest to fields, penetrating even higher up the slopes into what was then a reserve. If, he reasoned, one or more groups of the Grauer's gorillas that lived in these mountains could be habituated and become a tourist attraction, enough money might be earned to save the forest reserve. By 1970 his vision had become a reality. The reserve had become a national park, his efforts received international recognition, and he had shown that many tourists would clamber through dense forest to stand face to face with gorillas, not only paying fees to do so but also bringing much needed foreign exchange into the country. Censuses between 1976 and 1978 showed that only about 268 mountain gorillas remained in the Virunga Volcanoes.

In 1978 the New York Zoological Society—which had sponsored my work—sent Amy Vedder and Bill Weber to Rwanda to collaborate with the Office of Tourism and National Parks in gorilla conservation. Among their tasks were to study land-use practices of gorillas and humans, determine attitudes of local people toward the park, develop a tourist program, and initiate conservation education. Since half of the farmers at the edge of the park could not even describe a gorilla, a program to spread awareness of the need to protect the apes and their forest was rather urgent. The following year, a conservation consortium funded by several international organizations, including the Washington-based African Wildlife Foundation, initiated a complementary program. One emphasis was on park security, augmenting Dian Fossey's continuing war against poachers. Jean-Pierre von der Becke, a Belgian, took charge of anti-poaching. After the guard force had been increased, trained, and equipped, and a patrol system established, hundreds of snares were destroyed and within a few years fifteen gorillas were released unharmed from snares—though others still died.

Amy Vedder and Bill Weber visited schools throughout the country to talk to children about gorillas and the damage caused by deforestation. There were radio broadcasts about gorillas and conservation workshops for Rwandan science teachers. Posters and T-shirts advertised gorillas. A vehicle specially equipped for audio-visual presentations traveled from village to village. With the help of the Belgian Technical Assistance Program and the U.S. Peace Corps, and the full cooperation of Rwanda's Ministry of Education, conservation was added to the secondary-school curriculum. One report noted:

"This year we have taken out several groups of primary-school children and teachers to see gorillas and climb volcanoes. When these children told their families and relatives how well they had seen gorillas they were not believed—by the people who had lived all their lives near the Park."

The tourist program developed rapidly, from 1,352 paid visitors to the park in 1978 to 5,790 in 1984. In fact, by the mid-1980s, tourism had become the fourth largest earner of foreign exchange in Rwanda, and gorillas the country's main attraction. An economic argument on their behalf could now be made.

Conservation awareness increased markedly too. Should the park be converted to agriculture? When Bill Weber asked this question in 1980, fifty-one percent of the farmers answered "yes," but in 1984 only eighteen percent did so. What use does wildlife have, given that hunting is illegal? In 1980, forty-one percent could visualize some benefits, but in 1984, with a tourist program as evidence, sixty-three percent could. This integrated nationwide approach to gorilla conservation, imaginatively applied in cooperation with the Rwandan government, had within a short time a noticeable impact on the attitudes of the local people. And it is, of course, they who will ultimately determine the fate of the gorillas.

The initial costs of the program were small. To quote Alexander Harcourt: "The costs of these three programs to the consortium plus the Belgian Government was around $50,000 per year. The Office of National Parks paid a further $100,000. Each gorilla thus costs approximately $1,250 per year. To maintain them in a zoo would cost at least $1,500 annually."

Of course no dollar value can be placed on the benefits of saving a gorilla. The apes are the impetus for the conservation effort to protect the Virunga Volcanoes. Without gorillas the deep and fertile volcanic soil of the park would already be growing crops, without gorillas thousands of other plant and animal species would have been deprived of a home there. The Virunga range represents a mere half percent of the country's land but ten percent of the country's water catchment. Acting like a gigantic sponge, the forests hold water, storing it in the wet season and supplying it in the dry. If the forests were cut, perennial streams would disappear and the local economy suffer. Without gorillas the local people would already have cut their lifeline.

A 1981 census of gorillas gave a figure of about 254, a slight drop from the previous count. At first glance the lack of improvement was discouraging. However, the conservation effort had been directed only at Rwanda—and there gorilla reproduction was better than in previous years. By contrast, the Zaire side of the park had not improved. Sadly, Zaire and Rwanda have failed to cooperate in managing the Virunga Volcanoes. And in Uganda only nine square miles of reserve remain, most of them unsuitable gorilla habitat and none protected; no gorillas reside there permanently now.

In 1984, however, the Eastern Zaire Gorilla Conservation Project began in the Virunga Volcanoes, under the direction of Conrad and Rosalind Aveling who had previously worked in Rwanda. Their goal is to develop a conservation program for Zaire's mountain gorillas as well as Grauer's gorillas.

In the mid-1980s many individuals and organizations, aided by receptive governments, are helping the gorillas in the Virunga Volcanoes. The basic issues of human population growth and land use remain, and indeed

Carl Akeley's gorilla, shot in the Virungas in 1921. The Museum of Natural History, African Hall, New York, 1988. Photograph by Michael Nichols/Magnum. The only mountain gorillas in captivity are in the diorama displays of the museums of the world.

demand for tillable soil will increase year by year well into the next century, creating perhaps a "human wave powerful enough to wash the Virunga forest and all its gorillas away forever," as Amy Vedder noted. At least for the present, the preservation of the mountain gorilla is not just the vain pursuit of a romantic illusion; there is place for cautious optimism.

The mid-1980s also marked the end of an era in gorilla conservation. After eighteen years of devotion to the gorillas, Dian Fossey was slain by an unknown assailant on December 26, 1985 at her Karisoke Research Center. She lies buried in the Virunga Volcanoes, joining Carl Akeley, silent reminders to future generations of the love and dedication that gorillas inspire.

A census in 1986 revealed about 293 gorillas in the Virunga Volcanoes, the first definite increase in over a decade. Amy Vedder, who organized the census, wrote: "Moreover, the entire Virunga population could theoretically return to the 1960 level of 450 gorillas by the year 2010!" Dian Fossey would have been thrilled by this prognosis.

About twenty miles north of the Virunga Volcanoes, across hills that were once forested but are now wholly cultivated, is Uganda's Impenetrable Forest Reserve. Its 120 square miles of wooded ridges and lush valleys crowded with tree ferns are the home of the other mountain gorilla population. When I visited the reserve in 1960, I estimated that 120 to 180 gorillas lived there; later, in 1979, Alexander Harcourt calculated a population of 95 to 135. With neither survey precise, the extent of decline, if any, is unknown. Illegal logging, gold prospecting, and hunting has, however, increased in the past two decades. Conflict between gorillas and human beings is serious, so much so that the apes now avoid forest areas where disturbance is greatest. In the 1979 count, only thirty-one percent of the gorillas were subadult, less than eight years old. Since a figure of about forty percent is thought to represent a stable population, numbers in the Impenetrable Forest may well be declining. The status of these gorillas will soon become clear from the work of Thomas Butynski, who in 1986 with aid from the World Wildlife Fund established a research and conservation program in that forest. His project was timely, as these incidents illustrate:

In about 1979, two Austrians or Germans went on a trophy hunt and shot a silverback.

In 1983, two gorillas were killed by men with spears after raiding a banana plantation.

Gorillas were killed to capture infants at least twice in 1985. In the first instance the infant died on the way to Rwanda, and in the second the infant was successfully transferred to a buyer in Rwanda.

At present about 400 to 450 mountain gorillas survive in two isolated habitats which total 285 square miles. They can go nowhere else, adapt to no new way of life. The most elegant and powerful of our kin, they have almost disappeared, victims of humankind that has refused them their place in the forest. Few in number and highly localized, the mountain gorilla is a living blueprint for extinction. We now face the moral challenge and responsibility of providing a blueprint for the survival of these animals. It is sadly clear that the mountain gorilla will never be safe, that any success in protecting it will always be temporary. To assure the mountain ape a future will require a vigilance and dedication that are not measured in terms of projects or even years but in centuries, a commitment forevermore or until the sad day that man decides to discard this link to his past.

We can talk of protecting watersheds, the survival of ecosystems, and the economics of tourism, yet conservation has as much to do with ethical imperatives as with human poverty and the gorilla as a lucrative commodity. Societies must use their land with love and respect, they must change their loyalties and affection to guarantee all creatures the right to exist. And conservation also deals in emotion. When I look back on my time with the gorillas, I remember the magic of the animals, the beauty of a family patriarch, his silver saddle sparkling like morning frost. Such images are anchored in my heart. That in this supposedly enlightened age the mountain gorilla remains so vulnerable, its existence so wholly dependent on human whim, fills me with exasperation and indignation. Its loss would be a death in the family.

GEORGE B. SCHALLER

Gorilla

STRUGGLE FOR SURVIVAL
IN THE VIRUNGAS

The Virunga volcanoes, habitat of the mountain gorilla. Right to left, Gahinga (meaning "Hill of Cultivation"), Sabinyo ("Teeth of the Old One"), Mikeno ("The Naked One"), Visoke ("The Place Where the Herds Are Watered"), and Karisimbi ("Mountain of the Shell") seen from the summit of Muhavura (meaning "He Who Guides the Way"). The border between Rwanda (bottom), Uganda (top right), and Zaire (top left), runs through the mountain peaks.

The range of the gorillas is an expanse 40 miles long and some 6 to 12 miles in width. Two-thirds of the conservation area lies in Zaire in the Parc National des Virunga, while 37,000 acres in Rwanda is known as the Parc National des Volcans. The small northeastern corner in Uganda that is also a gorilla habitat is called the Kigezi Gorilla Sanctuary.

Rising to heights over 14,800 feet, the Virunga volcanoes encompass a rare and beautiful mountain forest ecosystem. Just prior to independence in 1958, the Belgian authorities permitted 7,000 hectares of park cleared for human settlement. Between 1969 and 1973 another 10,500 hectares were converted to agriculture under a joint development project managed by the Rwandan government and the European Economic Community. In 1979, 1,300 more hectares made room for settlers in the western sector. Finally, in 1986 some 300 hectares along the park border were returned to the park. The net effect was to halve the gorillas' habitat, from over 300 square kilometers to 150 today. Eliminated along with acres of habitat was a biologically diverse forest zone. A relic band remains between Gahinga and Sabinyo. There, shrubs and trees combine with numerous lianas, epiphytes and grasses. Bamboo (Arundinaria alpina) predominates along the park's lower boundary. At altitudes between 2,700 and 3,200 meters, Hagenia abbyssinica and Hypericum lanceolatum and abundant clearings of herbaceous vegetation flourish. Between 3,000 and 3,700 meters, giant heath formations extend down to forest zone along exposed rocky ridges. Senecio and giant Lobelia are farther up. Ground cover communities of blackberry, brambles, everlasting flowers, and shrubs mixed with lichens and grasses luxuriate, along with sedges, Sphagnum mosses and marshes. Above 4,000 meters even shrubs disappear, and only grasses, mosses and lichens prevail.

In the Virungas, more than sixty mammal and one hundred and eighty bird species have been identified. Several are considered to be endangered. Lions were last seen in 1943, but leopards, hyenas, jackals, civets, genets, serval cats, golden cats, and three kinds of mongoose persist. Today, the giant forest hog and the yellow-backed duiker are now thought to be extinct in the Virungas. But buffalo, bushbuck, black-fronted duiker, hyrax, dormouse, wild bush pig, and elephant can be found, along with eight types of primate, including the mountain gorilla, the golden monkey, the blue monkey, and two nocturnal primates: Perodicticus potto and Galago senegalensis.

It was Carl Linnaeus, the great zoologist, who in 1758 first devised the classification of animals that recognized the close relationship between human beings and apes. He accorded the highest ranking in the animal kingdom to man and the three great apes: the gorilla, orangutan, and chimpanzee—the only primates without tails. There are three subspecies of gorilla: the western lowland gorilla (Gorilla gorilla gorilla), most commonly seen in zoos and museum collections, of which at least ten thousand remain in the wild; the eastern lowland gorilla (Gorilla gorilla graueri) with less than two dozen in captivity and perhaps a few thousand in the wild; and lastly, the mountain gorilla (Gorilla gorilla beringei), with not a single animal in captivity and fewer than 300 surviving in the forest of the Virungas.

Mrithri in the bamboo.

Ndume, silverback leader of Group 11. While still a blackback (a young male gorilla), Ndume had to assume command of the group after the silverback died. Ndume, along with two other members of his group, was the victim of snares set by poachers for antelope or other game, and has only one hand.

Ziz, silverback leader of Group 5, with his family at midday watering hole.

Rugabo (meaning "chief" in Zairois dialect), the silverback leader of the Rugabo Group in Zaire. Adult male gorillas turn silver when they reach maturity, generally between the ages of 13 and 15 (though they have been know to turn silver prematurely if they assume leadership early). The silverback is the undisputed leader of a gorilla group, determining which direction the group travels, how far they wander and where they nest for the night. When the safety of the group is threatened, the silverback will remain to challenge the attacker while his family flees to safety.

Kwita, female infant, 2½ years old, from Group 9.

The family of Bidele. In 1969 Bidele was given two hectares of land at the foot of the volcanoes, in what was then parkland, on condition that he would farm a percentage of that land in pyrethrum. Today Bidele and his two wives have a family of twelve, including two older sons, nearly adult, whose only hope of having their own farms is if Bidele subdivides his small plot. Each year, an estimated 27,000 new families in Rwanda need land.

Portraits of Hutu farmers living in the townships surrounding the Parc des Volcans (Overleaf).

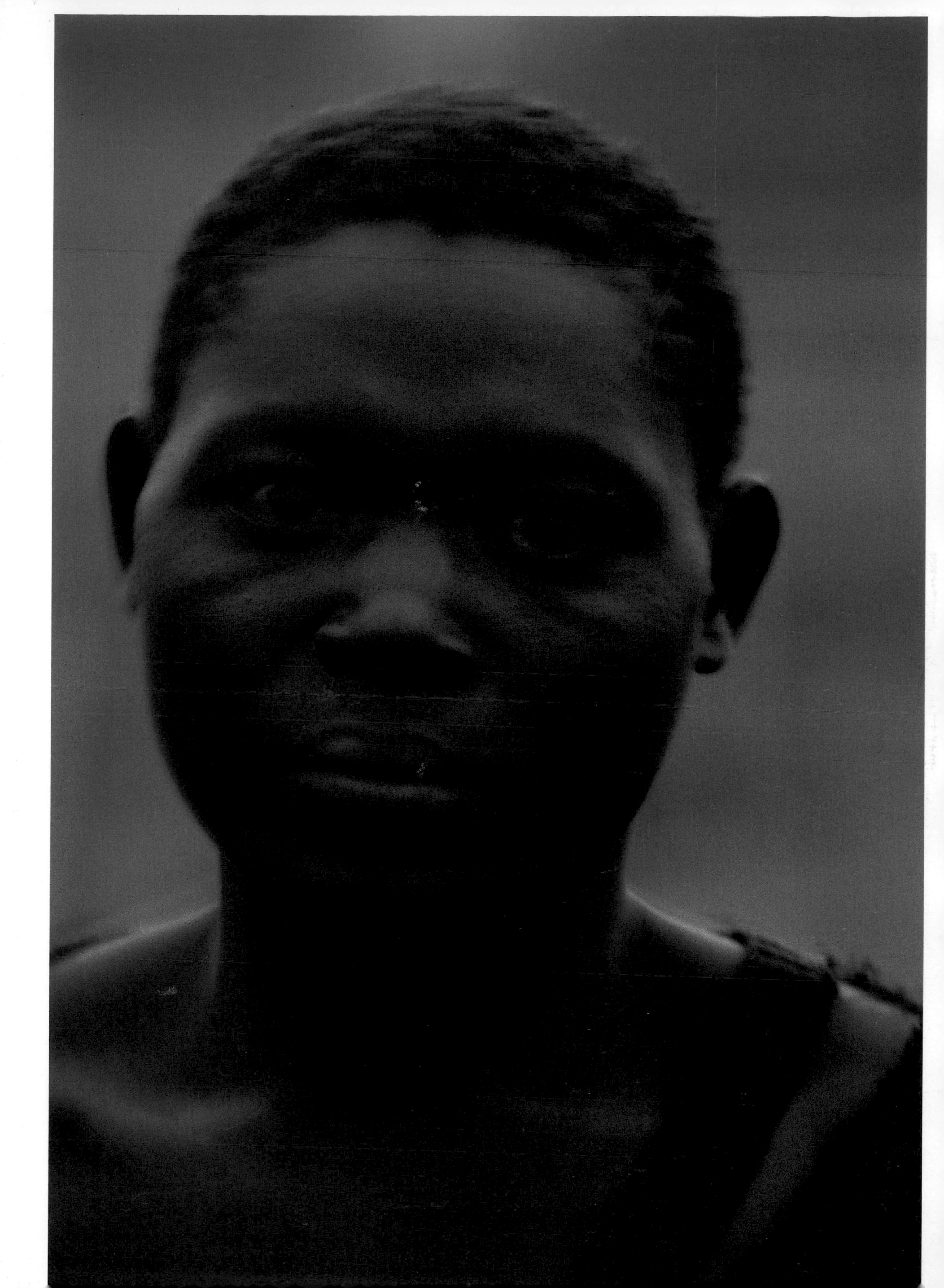

The Hutu were the largest of the three main tribes in Rwanda (The Tutsi and the Twa are the other two tribes) and comprise nearly 90% of the population. Mainly an agricultural people, the Hutu have adopted some of the habits of their former feudal warrior-lords, the Tutsi, traditional cattle-herders, who used to graze their herds in the park. In Rwanda, the lyre-horned Ankole cattle are a symbol of wealth and status in the community, and are greatly prized. The tall, frail Tutsi (opposite page), barely 10% of the population, were toppled from centuries of power in a bloody civil war in the 1960s. Over 200,000 Tutsi fled the country at the time rather than risk reprisals. The Twa, above, only 1% of the country, are a pygmoid people, and the original tribe of hunter-gatherers in the mountain forest.

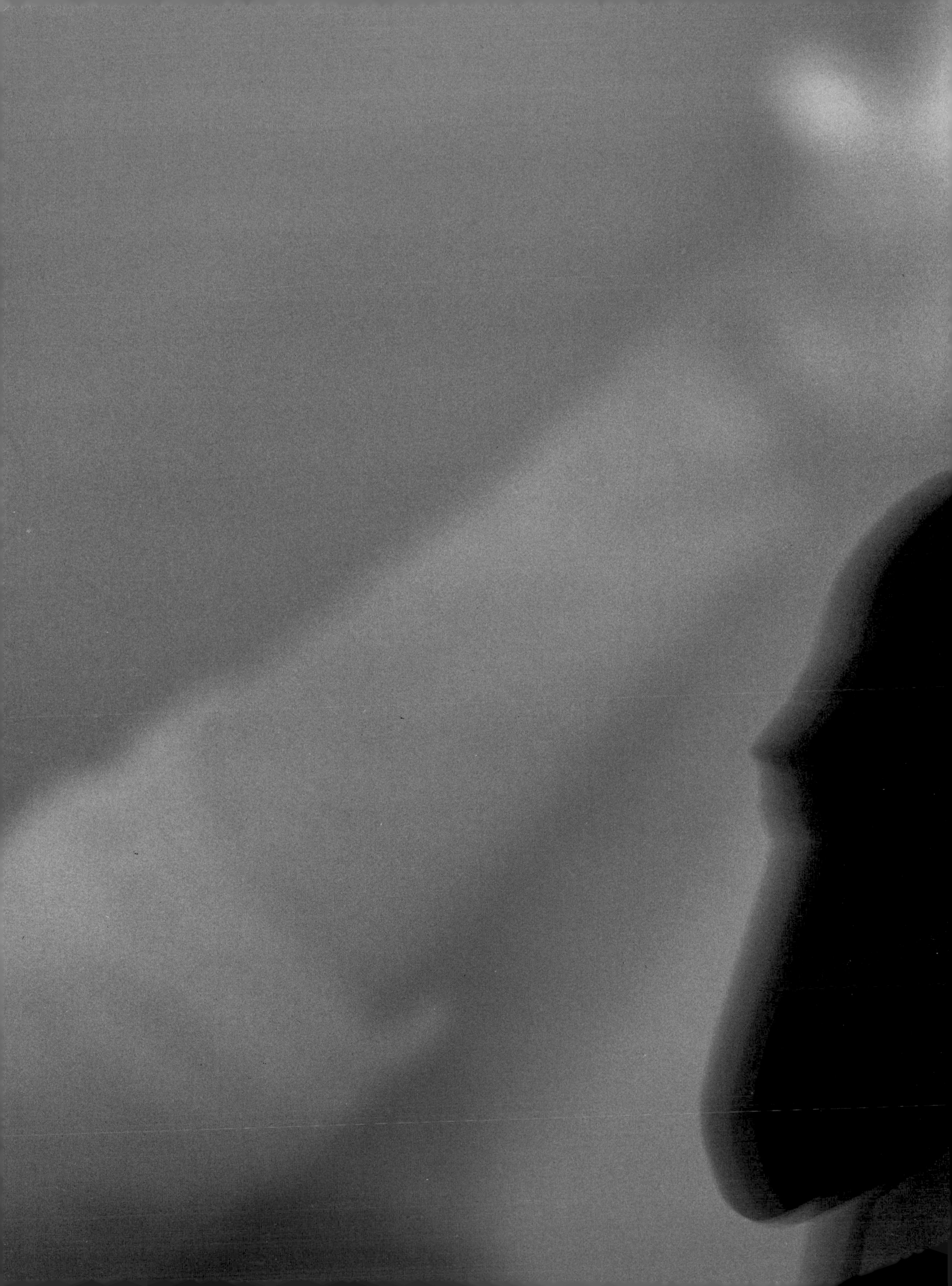

Giant heather silhouetted at sunrise on Mt. Sabinyo, with Mt. Gahinga and Mt. Muhavura in the background.

Ziz hooting. Though generally quiet, the gorilla can make twenty-five distinct sounds. Hooting, a sound that can carry for a mile through the forest, is usually exchanged between rival silverbacks and accompanied by chestbeating, strutting, vegetation-slapping, and other displays of vitality and strength. Some of the other vocalizations by gorillas include screams (signs of alarm or warning), belches (deep rumblings of contentment), sharp grunts (rebukes), high-pitched barks (curiosity), roars (aggression), and chuckles (indicating playfulness).

Mrithi cracking bamboo stalks. Mountain gorillas have impressive canines. The black stains on the portrait of Ziz (opposite), are from tartar. Gorillas consume vast amounts of vegetation. They obtain their water from moisture on the plants.

Bluff-charging Mrithi (right), slamming vegetation and hurling bamboo in a crescendo of noise and screams to frighten threatening intruders.

"Gathering food requires great strength, a quality also important in warding off predators. Strength plays a part, too, in reproductive competitions among males, in which fighting ability and the ability to protect females and infants is critical. Competition among females sometimes involves physical battles too, never as damaging as fights among males, but requiring strength nonetheless."
DAVID WATTS
former Director, Karisoke Research Center.

Sanduku, a blackback male in Peanut's all-male Group 8.

"How close they get surprised me—I was shocked. And their tolerance—a mother would allow an infant to come right up and touch you. At first I went by the Karisoke rule: contact is fine, but don't react if you can avoid it. After a while you begin to wonder—is it all right? When the infant approached I would back off. When an infant jumps on Ziz, he generally ignores him, but if he becomes too much of a pest Ziz knocks him off. If you go out of your way to ignore the animals that's not natural either. Umarava would strut up and act very disturbed when I avoided him. You should not tickle them or groom them and fondle them, though. I definitely don't encourage contact."

LORNA ANNESS, Karisoke researcher, notetaking in the rain.

"Relationships within the group are very important. For instance, one day Puck, an adult female, and her son Cantsbee, a young blackback, were settling in to look for nest sites when it started to rain. Puck sat next to Intwali, her two-year-old daughter, and Simba, in a good position well out of the downpour. Cantsbee came along, staring at Simba, and moving closer and closer. Simba gave a slow pig-grunt at Cantsbee, but then Puck pig-grunted at Simba while Cantsbee sidled even closer. Simba and Puck continued pig-grunting while Cantsbee slowly but surely pushed Simba out of the nest. He probably never would have tried to displace her if he had not known that Puck would back him up."

LORNA ANNESS

Hagenia abbysinnica is the most conspicuous tree in the Virungas. Its hoary limbs, growing to a height of seventy feet, support a profusion of other plants: orchids, ferns, mosses and lichens. The Hagenia is home to a variety of small mammals: dormouse and mongoose, hyrax, and squirrel, and its huge trunks provides shelter for the mountain gorilla and man alike. When in the park on a hunting sortie, poachers shelter in the hollowed boles of old Hagenia, in lean-tos called *ikiboogis*, and there they smoke marijuana around their night-fires.

"Evenings are quiet, except for the unearthly shrieking of the hyrax. That is the time when the solitude and the simplicity of life there closes in. It is not a way of life that many Westerners would desire, but it brings immense rewards to those who can endure the solitude without undue difficulty. It offers a chance to learn about the world and our place in it, both personally and collectively. The pleasure and privilege of knowing the gorillas and watching their lives is like nothing else in the world."

DAVID WATTS

"The question that interests me is why a female chooses one group over another—what makes a group desirable to a female. Is it because the group is good for the breeding and rearing of infants? The different status of females in a group has an effect on the social network of the infants in the group. There is no "typical" gorilla group, though Group 5 shows the extreme of social structure. It is an enormous group and there are many females that have large numbers of relatives within the group. But it still functions well. Hierarchy is more difficult to see because it is more subtle among gorillas than in other species—there is less competition over food. Maggie, for example, is a young female. Alone with females like Pandora or Picasso she is easily intimidated, even if she tries to act tough. But with her big sisters Puck and Tuck, she asserts herself. Dominant hierarchy is there. In all primatology, females have been the main interest because the groups are female-bonded. The theory is that females keep their relatives in the group to control food resources. But the gorillas are the exception to the rule, because the females can leave the group and seem to choose the group they join. And males often leave the group as well, at sexual maturity. It's a very special social structure. We try to remain passive when they come to us, we don't encourage them by grooming them. But these gorillas see the same people every day and become very used to us. Only a few come close: Maggie and Shinda are always interested when we wear bright colors, or T-shirts (which they like to chew). But they soon forget you and it is as if you weren't there."
PASCALE SICOTTE

"Doing science is like a jigsaw, but with pieces missing, and with only a little of the picture to work with. By comparing the gorilla piece with as many other pieces as possible, whose size and shape and color others have spent deciphering, we try to discover similarities and dif-

Pascale Sicotte, Karisoke researcher, studying female hierarchy within Group 5, alongside Ndatwa, a male juvenile.

ferences between them, and to work out the rules that govern their shape and the way they interlock. The gorillas has been a useful piece to research, partly because it is unusual. That makes it easier to see whether or not it fits into the existing picture and to understand why it does or does not fit. That understanding helps us and other biologists place other pieces in the pictures, or maybe even start a new picture. What picture, what scientific understanding have studies of the gorilla helped towards? One is why animals, and ultimately man, have the sort of societies that they do. Why do gorilla groups consist of small harems of a male and a few females? Why did nomadic aborigine groups number about 50 members with an equal ratio of men and women? Thanks in part to the gorilla studies, we know that the way resources are distributed is one crucial part of the answer. This knowledge in turn helps predictions of how animal and human societies might respond in today's rapidly
ALEXANDER HARCOURT,
Department of Applied Biology, Cambridge University, England

"At some point in the past, there were natural population controls among animals in the Virungas. Now, with the prevention of poaching, or the elimination by poachers of all the natural predators like leopards, the population of animals will increase—and that poses problems. At some point the park will have to cull and curb population increases in animals like the buffalo, who are pushing into the planted fields and destroying crops and boundary fences. But we have to know the dynamics of the place and all the animals before we make those decisions. On a much broader front, research has to go on, but we owe it to the people who live here to take on a broader attitude. Research training and conservation activities must address both park management and the people living at the borders of the park. As far as research is concerned, the longterm studies of the gorilla will continue and extend to cover other aspects: work on the all-male groups and lone males, studies of the soil and the climate, and particularly tropical botany. We need an ecosystems approach to the whole park."
ALAN GOODALL

Alan Goodall, Director of Karisoke, with Titus, the second silverback in Beetsme's group.

The graveyard at Karisoke, where alongside the graves of gorillas Digit and Uncle Bert, scientist Dian Fossey is also buried. Her flower-covered grave is marked by a photograph showing Fossey touching the gorilla Peanuts, a contact that marked the culmination of the process of habituation. Fossey established Karisoke as a research station in the Virungas in 1966 and in 1970, when *National Geographic* magazine published the first of three articles on her research. By the late 1970s Fossey's scientific work had taken second place to anti-poaching activities, when in 1977 one of her favorite animals, Digit, was found slain. His much-publicized death led to the founding of the Digit Fund (now under the auspices of the Morris Animal Foundation) and the formation of the Mountain Gorilla Project by a consortium of international conservation organizations, which proposed a three-part plan of education, anti-poaching, and tourism as a way of protecting the gorillas. Fossey was vehemently opposed to tourism, fearing that excessive contact would inevitably create grave new problems for the gorillas. Photograph: in insert by Bob Campbell.

The guides and trackers of the Karisoke Research Center, left to right: Nshogoza Fidele, Nemeye Claver, Munyanshoza Leonard, Elias Rukera, Nsezeyintwali Gabriel, Bizumuremyi Jean-Bosco, Munyanganga Kana, Kanyarugano Leonidas; and also (not pictured here), Banyangandora Antoin, and Nkeramugaba Celestln.

'In June, 1969 I started working at Karisoke as a porter. When I first arrived I was sixteen years old. It was my first job and I had never been in the forest before. Dian and Vatiri showed me gorilla tracking. I have been here many years now and have seen the researchers come and go. I was here when there were still cattle in the park. One of the problems was that the gorillas wanted to come down during the early part of the dry season and there were many cattle. The President has impressed upon us the importance of keeping the number of cattle down. They are our wealth and the people get milk and oil from them, but with fewer cattle we can better preserve the park. Because of the words of the President, now there is much more awareness of the killing of animals and the poaching problem. I am thirty-five years old and I think forty-five will be enough. This is my home, my work, my pay. But I do not work just for the pay. All these *Muzungus* come here to study gorillas and the gorillas are very few; if we were not guarding them, they would disappear. The future is this: the conservation of the animals here will make sure there is work for their children and our children."

NEMEYE CLAVER
("Big Nemeye")

The wall of Dian Fossey's cabin, with its collage of gorilla photographs.

"The other Rwandans think we have poisoned the gorillas or drugged them so that people get close to them. They think it is very hard work and dangerous. The fact we keep returning is proof that the gorillas have not eaten us. Now the Rwandans know that the gorillas don't eat meat, but still they are surprised to see us uninjured, and therefore have a lot of respect for the work and are happy that people get work because of the gorillas."

LEONIDAS ZIMULINDA

Sunrise over the Virungas. The cloud-covered peak of Mt. Mikeno and the darkened slopes of Mt. Visoke.

Porters light campfires on the slopes of Mt. Karisimbi, at an altitude of 12,000 feet. Through the dense and softly-textured vegetation of the mountain slopes, the muddy trail tilts upwards at a 45-degree angle, through stinging nettles and blackberry bushes, and the cold and mist of a forest where the annual rainfall exceeds seventy-two inches.

Shabani, a porter, climbing Mt. Visoke above Lake Ngezi, through mountain gorilla habitat.

Moonlight and shooting stars over the crater lake at Mt. Visoke's summit, 3,711 meters high. Mt. Mikeno to the left, Mt. Karisimbi to the right.

The three-horned chameleon (Chameleo johnstoni) exists on the Zaire side of the Virunga volcanoes. The chameleon figures prominently in African folklore and mythology and, in spite of its slow movements and seeming slow-wittedness, is often linked with errands of great importance and issues of mortality. In one Pygmy myth, set in time before man appeared and before water existed on earth, the chameleon was climbing a tree and heard whispering inside the trunk. The little lizard bored into the tree and water poured forth. The gods were so taken with this discovery that they created man to use the water. In another legend, a bitter tale of drought suffocating the land, the sun glowed in a perpetual fiery haze across Africa. The bones of dead animals crackled in the heat and even the locusts stopped coming. The elders of the village charged the chameleon with the sacred trust of reaching the gods with their plea for rain to save the earth, but by the time the slow little messenger arrived, the people had perished.

Inside the smoking crater of Nyamulagira, one of the two active volcanoes (along with Nyiragongo) in the Virunga mountain chain. Gorillas are not found on these two volcanoes.

Sanduku, a young male from Peanut's group, navigating the mossy branches of a Hagenia tree in search of a nesting site. Bending the leafy stalks of Lobelia and Senecio trees into a padded cushion of foliage, gorillas construct nests every evening. The nests are usually found among the branches or hollows of trees, or on sheltered knolls on the ground offering vantage points of the surrounding terrain.

Infant gorillas share a nest with their mother until about the age of three, or until the mother again gives birth.

"We walked through the forest in a pattern we knew would sweep the areas frequented by the gorillas in the park, moving from west to east, following trails no more than seven days old (determined by how the plants have been trampled). We search a trail which includes at least three night-nests, then stop and count the number of nests at each site. Once the nest is located, we measure the dung left in the nest, which gives an excellent idea of the composition of a family. For younger animals, we have to examine the nest carefully. Usually they nest with the mother, but on occasion, when the mother has transferred out of the group, they have been known to nest with the father. (And infants of less than six months have no dung and generally go

uncounted.) In groups not visited regularly, we try to contact the group, and at least attempt to observe the silverback, to get a sketch of his nose-print. That way, we can distinguish any accidental repetitions in the count, in case two different census workers happened upon the same group. The impetus for this census was the debate about the effect of tourism on the gorillas; one way to monitor the tourist program's success or failure was through the census. We split the data into tourist and research groups and found that a greater percentage of young survived in the tourist groups than in any other. There was, in fact no foundation to the idea that contact with the gorillas was inhibiting reproduction. The population count, as estimated by Schaller in 1959, was substantially reduced by poaching and loss of habitat in the early 1960s and early 1970's, as noted by Harcourt and Groom in 1973. Bill Weber's 1978 census, and Harcourt's follow-up in 1981, showed a lessening of the decline, especially in Rwanda. Our last census, with its finding of a positive upturn in both the total population and the percentage of young, is an index of optimism for the future of the mountain gorilla."

AMY VEDDER,
ecologist

Kubinya from Peanut's group, resting in the crotch of the Hagenia.

Mrithi among the bamboo. Primarily vegetarian, the gorillas feed on tender white shoots of bamboo, which are seasonal, and favorite foods such as thistles, wild celery, and other herbs, fruit, and leaves of many trees, such as those of Musanga, Myrianthus and Ficus.

The hand of Pandora, a female in Group 5, whose hand was thought to be deformed by a wire snare when she was young, or possibly by disease. In recent years, the daily monitoring of gorillas by researchers and guards, and quick location and removal of gorillas from snares, has significantly reduced the number of incidents of lost gorilla limbs.

A metal snare used to poach duiker, a small antelope. The snare is tension-sprung through contact with a bamboo pole, and pulls up to tighten on the leg of the duiker, lifting him above ground. There the animal hangs, suffering a slow and agonizing death, until the poachers return, often days later. The same snare often traps gorillas, cutting deep into their flesh and causing infection and possible loss of limbs, or even death.

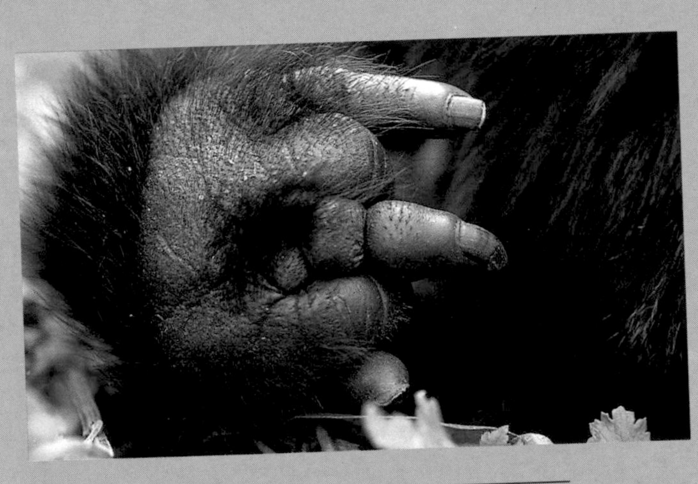

"There are two kinds of poachers: one hunts to feed his family, and another hunts to sell meat to hotels and rich Rwandans, to sell elephants for ivory, antelope for skins, and gorillas for their skulls. But now, the minimum penalty for killing a gorilla is five years in jail. We still find snares, but they are for duiker, and you cannot stop that completely. The main problem is that the gorillas are trapped in the snares, too. We are the first to intervene if a gorilla does get caught; we decide whether to anesthetize the gorilla and remove the wire."
JEAN-PIERRE VON DER BECKE
former Director of the Mountain Gorilla Project.

Sylvestre, an ORTPN (Office Rwandais du Tourisme des Parcs Nationaux) guide, holds a buffalo snare, which is buried out of sight in a mudhole and traps the buffalo by means of a bamboo foot-harness connected by a line to a large log. As the buffalo moves, the line tightens, dragging the log between the trees, and rendering the buffalo immobile. His distress calls attract the poachers, who descend for the kill.

Duiker carcass caught in a snare; similar snares often trap gorillas.

"The idea of an education project was to sensitize the Rwandan population, especially the rural population around the park, to the importance of the park for the economy and the lives of the people themselves. It had three objectives: to promote a better understanding of the park's benefit to the local people; to create a favorable sentiment toward the park in the local villages, making cooperation between park and civil authorities easier; and lastly, to sensitize the people to regional environmental problems and to the means to reduce these problems. We began by contacting, through a letter from the director of ORTPN, the important leaders of town prefectures, the directors of secondary schools, and the education minister and inspectors teaching in primary schools. We held meetings in the townships and answered questions afterward. We have begun a library with scientific documentation on the park. We are publishing articles in national and international papers and jour-

nals. We are also producing information sheets on specific problems relating to the park: poaching, woodcutting, and much more. The forest is important as a watershed, as a store for flora and fauna, and we must protect it against fire, woodcutting, animal poaching, and other encroachments. In the schools our message is: this is our childrens' responsibility for the future."
FRANÇOIS MINANI,
Education Coordinator,
Mountain Gorilla Project

Schoolgirls watching films on gorillas. Educational films on the gorillas are often shown at local villages.

"If this were the best of worlds, the sheer natural beauty of the animal would have sufficed as argument. But the Africans didn't even know whether they were beautiful or impressive—they'd never seen them. When we did surveys in the late 1970s, more than half the local population could not provide a single adjective of description. They lived in the shadow of these volcanoes and had no idea what the gorillas were like. So, they would flock to the films we showed, up to 3,000 at a time from a village area, and the response was incredible. They saw how human this animal was. They saw a family unit; they saw a lot of maternal care—nursing, cleaning, playing, feeding; they saw paternal affection. These films did not show King Kong bursting through the bushes."
BILL WEBER,
Assistant Director of Conservation, The New York Zoological Society

"I was looking for a job and heard that they were looking for guides to the gorillas. I had seen pictures of gorillas but was afraid. The first day my heart was beating very fast, but we stayed an hour, and by the end of the time I was happy and liked the gorillas. After one year of learning to track, I took an exam on the gorillas. I had to know their names, their faces; I had to know the plants in the forest. I had to track them for four days without help, and I had to know how to give instruction to tourists. We are careful not to let visitors touch the gorillas because of human disease. We show the visitors how to act so the gorillas are not disturbed. To have too many people would make the gorillas aggressive. Now I have had three years here in the forest."
JEAN-BAPTISTE

Porter carries coolers filled with champagne up the steep slopes of the mountains for a group of visiting tourists. The income generated from tourism is the third largest source of foreign revenue in Rwanda.

Tourists observing Group 9. A visit to the gorillas costs about 120 US dollars in Rwanda ($40 in Zaire); safari costs vary, beginning at around $3,000. A maximum of six tourists visit one group for one hour of observation, sometimes walking and hiking for hours through dense vegetation and stinging nettles before they first glimpse the gorillas. In just two years, the Zaire tourism program has become the biggest income-producer in the Zaire park system and effectively pays for the guides and guards in all Zaire parks. In this sense, protecting gorillas also protects the white rhino and the okapi, two other endangered species found in Zaire.

Conrad Aveling with tourists observing the Rugabo group in Zaire. Aveling, along with his wife, Rosalind, worked with the Mountain Gorilla Project in Rwanda, continuing the three-part approach toward conservation: anti-poaching patrols, economic incentives, and education. They have set up a similar program in Zaire, where more than half the gorilla population roams.

Front wall of a *pombe* (banana beer) bar in Zaire.

"The education project was a stone thrown into a pool, where it sank due to funding problems; but the ripples continued. The image of the mountain gorilla began to turn up everywhere. Now the gorilla has become the de facto symbol of the country, something which it is proud of. The gorilla is on the national currency; popular songs have been written about their plight; car bumper stickers display the gorilla's picture, and some hotels even offer soap bars embossed with the gorilla's image! This represents the beginning of probably the most important development of all—a change in attitude. It's uncertain, it's time-consuming, it's definitely the most difficult—but it's the most critical change, because in the long run if people value something, they'll probably protect it."
BILL WEBER

The Rwandan 1,000-franc note bears the image of a mother and baby gorilla.

"I have been a gorilla guide for ten years. I started as a porter then learned to be a guide. At that time the gorillas were not habituated. I learned to stand still when the gorillas charge and not to run—it is dangerous to run. I learned when the little ones come close to you the silverback gets mad; then he hits the ground hard and grabs the babies and hits the females. If you follow, the silverback will scream and make them leave. The gorillas know the faces of the people who follow them: Old Zahabu (in Group 13) screams when she sees tourists, but when she sees us, she just goes on eating. For me the most interesting thing is to see the gorillas very close, to see the silverback protect his children and hold them or let his children play on his back. The Rwandans see that the gorillas are more interesting to the tourists than the Rwandans themselves, but they like the gorillas because they don't destroy the agriculture like the buffalo do, and because the gorillas live in the park, and the park brings work to guides, guards, and porters—work to many people."
LEONIDAS ZIMULINDA

Children demonstrate growing awareness of the gorillas. This drawing by a Hutu boy shows tourists with cameras observing the gorillas.

Hutu boys with helicopters made from vegetables and bamboo. They see helicopters or planes on the rare occasions when the President of Rwanda visits the park.

Family portrait of Francois Ngaugwanayo, pastor of the Protestant church at Visoke, with his wife and five children, each born a year apart. Rwanda, known as the "Land of a Thousand Hills," has the highest population density in Africa, nearly five hundred people per square mile, and the highest density of rural population in the world. Competition for available land is great. The national park established in 1929 has been truncated three times: in 1958, 1973 and 1979, when 38.5 square miles of parkland was taken from the gorilla habitat in Rwanda. The pressure this ever-expanding population creates on the government, who is forced to find ways to feed its people, has important consequences for the future of the gorillas.

An agricultural people, the Hutu produce potatoes, corn, beans, coffee, tea, and sorghum and other subsistence crops.

"Rwanda is overpopulated in relation to its resource base. For all the good efforts of the people there, it's reaching a point where malnutrition or undernutrition is widespread and will become more generic in years to come. The area around the park is one of the most densely populated areas in the country, because the soil there is most fertile, with all that rich volcanic earth. By the year 2000, the country will have ten million people. Think about that, see that every square inch that isn't park land is already under cultivation, and that the sons of farmers cannot find land—or don't want to farm but at least be part of a modern world. But there is no place for them, because there isn't enough of a modern sector in the economy. How will they be fed, how will they live? It is hard not to think that before too long, people will go in and cut what is left of the forests in the Virungas and Nyungwe and use it to keep themselves going just a little longer."

DAVID WATTS

Hutu family harvesting pyrethrum. Potato fields in the rich volcanic soil of the Virunga foothills.

"The pyrethrum project grew out of the reaction against DDT in the West. People were looking for a substitute pesticide that would be less toxic, and the most promising candidate was the natural insecticide pyrethrin, distilled from the flowers of the pyrethrum plant, which can grow only at high elevations. The EEC came to Rwanda looking for a source. Kenya was already growing the plant, mostly for domestic use, and quickly expanded production to meet international needs. When for a while that demand could not be met, the case was bought and sold in Rwanda that this could be a major income-earner. Seeing no particular benefits from their forest, Rwandans converted 10,000 hectares of the park to pyrethrum development. They divided the land into two-hectare plots and allowed 5,000 families to farm there on condition that 40% of their land produce pyrethrum. In five years, the bottom fell out of the market; the West had synthesized other pesticides. This sad story, unfortunately, has happened to so many natural products of the tropics: someone else develops an alternative, a more secure, or better-refined supply, and then cuts off the original source. Now the fields of pyrethrum are down to 10% and people have turned to planting potatoes on the land. But the end result is the same—the gorillas have lost 40% of their habitat."

BILL WEBER

Bahutu woman hoeing the *shambas* (fields), in the shadow of Mt. Visoke.

Bahutu grandmother with her children and grandchildren; Bahutu boy, Gaston, son of Bidele.

"Long ago, Rwandans thought that volcanoes brought the evil eye; no one named a volcano before breakfast because doing so was supposed to unleash bad luck all day long. Others, seeing fire and smoke bursting from the peaks, thought the volcanoes were dwelling places for spirits that had to be cared for by the living. They said, "The spirits are lighting fires to warm themselves, making fires because they are cold." I join with those who thought this in explaining what is true, so that we can come together and protect and preserve the spirits' dwelling place, the place of our ancestors. In the home of the spirits, their watchman never sleeps at night. Even in the daytime he never hides. His stature and his massive form have attracted foreigners from many countries to come to pay their respects. He is an important emissary, bringing other lands to us. . . . This game preserve is a beautiful pearl . . . We must protect this park, and its animals. Let it be our pride. Let us remember the promise of the President of our country, and know that Rwanda will be built with her children's hands."
LAURENT HABIYAREMY,
Director of ORTPN

"I was born near the volcanoes, just fifteen kilometers from the boundary. The people around the park understand that the forest is for the gorillas, but there are other benefits, too: the forest is a watershed that brings the rain and controls erosion; the forest is a great sponge that catches the water in the rainy season and slowly releases it to the farmlands below, even in the dryest times. While we need more land to feed the people, and there is not enough land for parents to give their children, we understand that we are responsible for protecting the land and its future, and our future."
SEBIGOLI ELII,
Education Officer, Mountain Gorilla Project

Queuing for classes at the Visoke schoolhouse and church, the girls wear blue, and the boys tan, uniforms, often their only clothing. At the road's end, the trail leads up to the Karisoke Research Center, nestled at an elevation of 10,000 feet (Overleaf).

Transporting beans and potatoes to the market (Overleaf).

Beetsme with Jenny, a juvenile female, taking a midday siesta with other members of his family of nine, including Titus, the young silverback, Papoose, and Fuddle; the females Mwingu, Shangaza, Ginseng, and infant Kyryama.

Kudatinya and her infant son, Gashungura, 2½ years old, from Group 9.

The riotous offspring of Ziz play on their father's back.

Newborns are tiny, weighing only four to five pounds, and are cared for at all times by their mothers, one of whom is seen here clutching the baby to her chest. By the age of two, infants show a marked increase in movement, their upper and lower middle teeth have appeared, and they reach and chew on branches and vines. The rate of development is roughly twice as fast as that of a human baby.

Four-month-old Ugenda, a male infant, clutching his mother, Walanza, from Group 5.

"Ziz had the good fortune to mature in a group where his father was the dominant silverback (Beethoven) and to stay with that group and mate with several females (his half-sisters) born in the group. Father-daughter matings do not occur, and rarely does a male shows sexual interest in his daughters, but sibling matings seem normal. By 1984, Ziz became the only breeding male in his group, and a remarkably successful one, attracting a great influx of females who became Ziz's mates (though some copulated with Pablo, who was reaching maturity at the time). Ziz is the largest male gorilla ever seen, but he is also socially mature for a silverback so young (he is 17). Success in building a group takes more than physical prowess; it requires subtle skills to develop relationships with females and young. Personality is key."
DAVID WATTS

Kampanga, a five-year-old female from Group 13 (right). Effie, from Group 5 (below), sunning after a hard rain, arms akimbo.

Pablo, a silverback in Group 5. "Pablo and Ziz are half-brothers and sexually competitive. Researchers speculate he is tolerated by Ziz as an ally, perhaps because of the unusually large size of the group, which has twenty-nine members. Over the last few years, Pablo had the normal difficult life of the adolescent, trying to make the females take him seriously, showing them how big and strong he is, while avoiding Ziz, who is wary of any challenge to his authority. Ziz is often intolerant of Pablo and may yet force him to leave the group. Pablo's hard work cultivating social relationships within the group, grooming others constantly, becoming the special friend of several of the females, has created almost a group within the group, and if he did leave, they might go with him. He is both the pretender and just an overgrown, mischievous kid."
DAVID WATTS

Mountain gorillas have strong attachments to their group, and individuals often stay together for years, though shifts in membership do occur. Females usually transfer to other groups, especially if the silverback is their father and there is no other suitable mate within the group. Adult males generally leave after reaching sexual maturity, since mating with females of the same group is not permitted by the silverback.

"There was still poaching of gorillas in Rwanda until 1983. It was almost a yearly occurrence. To get a baby gorilla they often had to kill the silverback, the mother, and whoever else came to the aid of the baby, so the end result was two or three gorillas killed, the baby removed from the population, and in most cases, left to die in the hands of the poachers because they would do this on speculation and would then go about looking for buyers."
MARK CONDIOTTI,
Assistant Director,
Mountain Gorilla Project

"Poachers on the Zaire side are heavily armed. They have automatic weapons and there are shootouts, where guards and poachers alike are killed. But they are after elephants, not gorillas. Gorilla poaching seems to have stopped. Rwandans put an end to that with all the publicity after the deaths of Digit and Uncle Bert. The word spread that there was no market for gorillas. Back in the States and in Europe, gorilla consciousness soared. *Muzungus* [whites] are no longer interested in infants and artifacts—and if anyone tries, they'll get caught. But of course animal traps also catch gorillas, and there are plenty of mutilated gorillas. You still have poachers in the forest after duiker, bushbuck, and elephant. For example, the guards at Bukima heard shooting and went out to investigate, except for one tracker who went to follow the gorillas as usual. They came across a recently-killed elephant with tusks removed, but no poachers. Meanwhile, the tracker reached the gorillas, and what did he see but the poachers doing a bit of tourism on their way home. Old Rafiki, the blackback, was performing gymnastics in front of them, as the poachers sat watching with their AK-47s on their knees and their two freshly-removed elephant tusks beside them. The tracker fled and later turned in his resignation, saying, "This is too dangerous for me." Luckily, he was persuaded to rejoin, because he is a very good tracker."
CONRAD AVELING,
Director, Zaire Gorilla Project

Gorilla hand. When first visiting Rwanda, Dian Fossey's cook, awestruck at seeing his first gorilla, whispered in Swahili, "Surely, God, they are my kin." The four fingers of the hand of a silverback have a width of six inches. The foot print of a silverback is as long as twelve inches and much broader than a human foot.

"Primates are the most social mammals on earth and their study has been central to our understanding of how social behaviour evolved. After decades of research, it is now known that matrilineal kinship is a primary force in that evolution. In most monkey populations, females remain for life in the group in which they were born, whereas males usually leave at around adolescence. Thus, the core of social groups are clusters of related females—mothers, daughters, sisters, grandmothers. Gorillas are among the very few species in which females habitually leave their close kin and join other groups. Study of the gorillas has led to an understanding of the ecological and social forces that might shape male-female bonds, bonds so central to our own society."
KELLY STEWART,
Scientist and former
Karisoke researcher

Portrait of Beetsme, silverback of the group in Rwanda that bears his name.

Chonin, silverback leader of Group 9. Gorillas are the largest of the great apes, with an adult male weighing as much as 400 pounds and reaching six feet in height. Their life expectancy is thirty-five to forty years of age.

Mawingo, a juvenile female from Beetsme's group, sleeping.

"Sometimes they sleep, not at night, but during the day. When we are visiting they keep their eyes open, always watching. They may not be looking directly at us but they are always watching. It's very rare that they would sleep in front of you; that is why we have the rule limiting hours to visit, because they need time to relax without being wary. They eat for about an hour, digest, and then, especially around midday when the sun comes out, they lay down in the grass and relax."

MARK CONDIOTTI

CHRONOLOGY

1846 Savage and Wilson, missionaries in Gabon, send skulls to anatomists Wyman and Owen and are therefore credited with discovering *Gorilla gorilla gorilla*.

1856 Paul Du Chaillu becomes the first white man to shoot a gorilla in West Africa. In 1861, he publishes *Explorations and Adventures in Equatorial Africa*, popularizing the gorilla with exaggerated descriptions of its ferocity.

1861 Speke and Grant, searching for the source of the Nile, travel east of the Virungas, becoming the first Europeans to penetrate the region. They hear of "man-like monsters, who could not converse with men" inhabiting the mountains to the west.

1902 Captain Oscar Von Beringe, a German officer attempting to climb Mount Sabinyo, shoots two apes. Von Beringe sends the skeleton of one of the apes to the German anatomist, Matschie, and the mountain gorilla is named *Gorilla gorilla beringei*.

1902–1925 Museum collectors shoot fifty-four gorillas in the Virunga volcanoes.

1921 Carl Akeley shoots five gorillas for the American Museum of Natural History, New York. Akeley was so impressed with the gorilla and the mountains that he urged the Belgian government to set aside a permanent sanctuary for the animals where they could live in peace and be studied by scientists.

April 21, 1925 Albert National Park is established by King Albert I of Belgium, protecting the area around Mikeno, Karisimbi and Visoke volcanoes. It is enlarged on July 9, 1929 to include the entire chain of the Virunga volcanoes.

1926 Akeley returns to the Virungas to study (*not* shoot) gorillas. He dies of malaria and is buried at Kabara Meadow below Mount Mikeno.

1959–1960 George B. Schaller conducts his landmark study of gorillas based on Mt. Mikeno from a base camp in Kabara Meadow. This culminates with the publication of the scientific monograph, *The Mountain Gorilla*, in 1963, followed by the popular *Year of the Gorilla*, in 1964.

1962 Rwanda gains independence and the Hutu tribe are established as political rulers. In the following years, thousands of Tutsi, the former tribal power, are killed, with tens of thousands more forced into exile. Albert Park is split into the Parc des Volcans (Rwanda) and the Parc des Virunga (Zaire).

1967 Encouraged by Louis Leakey, Dian Fossey begins her study of mountain gorillas, also at Kabara. After six months, unrest in the Congo (now Zaire) forces her to move her study area to Rwanda. She sets up camp between Mount Visoke and Mount Karisimbi, calling the new camp Karisoke.

1969 Forty percent of the total area in the Parc des Volcans, Rwanda, is eliminated by conversion to farmland, in a European-sponsored scheme to grow pyrethrum.

"Coco" and "Pucker," gorilla infants confiscated from poachers, are nursed back to health by Dian Fossey at Karisoke, only to be later exported to the Cologne Zoo by the Rwandan Park Service. They become the only mountain gorillas in captivity (Nine years later, both gorillas die.)

1970 "Peanuts," a young blackback male, touches Dian Fossey, and becomes the first gorilla living in the wild ever to reach for and touch a human being in a friendly manner. That moment of overt acceptance marked the culmination of the mountain gorilla's habituation to human beings.

1977 Group 4, led by Uncle Bert, is attacked by poachers and Digit, the young silverback, is killed while protecting his family, his head and hands hacked off.

1978 Uncle Bert and Macho are killed in a poachers' attack on Group 4, in which Uncle Bert is also decapitated. Kueli, Macho's infant son, is wounded and dies three months later.

1979 Spurred by publicity over the gorilla murders, and guided by recommendations by New York Zoological Society researchers William Weber and Amy Vedder, a consortium of international conservation organizations forms the Mountain Gorilla Project, with the idea to produce revenue for the Parc des Volcans through controlled gorilla tourism, and to further protect the gorillas and their habitat through education and anti-poaching programs.

1981 The census finds 254 mountain gorillas in the Virungas—a marked decline from an estimated 400–500 gorillas counted during Schaller's study.

1984 Conrad and Rosalind Aveling start the Zaire Gorilla Conservation Project, funded by the Frankfurt Zoological Society, extending protection to over two-thirds of mountain gorilla habitat.

Over 5,790 visitors attend the Parc des Volcans in Rwanda, an increase from 1,352 in 1978, the year before the Mountain Gorilla Project was started. Gorilla tourism becomes the fourth highest source of foreign currency.

1985 Dian Fossey is brutally murdered in her cabin at Karisoke. She is buried in the gorilla cemetery behind her house. David Watts takes over as director of Karisoke, continuing the research project.

1987 The census, led by Amy Vedder and Conrad Aveling, censused 293 mountain gorillas, an improvement over 1981 figures. They find a marked increase in immatures, which is a strong indication of a healthy population.

1988 The Morris Animal Foundation and Veterinarian Jim Foster open the Virunga Veterinary Center for the care of mountain gorillas and other animals, and to help in the research at Karisoke.

Twenty infants are born to research and tourist gorilla groups in Zaire and Rwanda.

Rugabo leads his family off down a trail for early evening foraging before nesting for the night.

"If you want to talk about the future, talk about the future of the human race. In the next twenty years there are worse things that might happen to the whole planet than could happen to the mountain gorillas. With continued, thoughtful protection, the population of gorillas could double in the years ahead; the amount of land can sustain much more than the present population. But a great deal depends on man's relationship to them."
MARK CONDIOTTI

BIBLIOGRAPHY

Akeley, C. 1923. *In Brightest Africa*. New York: Garden City.

Akeley, M.J. 1929. *Carl Akeley's Africa*. New York: Blue Ribbon Press.

Akeley, C. E., and Akeley, M. L. J. 1932. *Lions, Gorillas and Their Neighbors*. New York: Dodd, Mead.

Aveling, C. & Harcourt, A.H. 1984. A Census of the Virungas Gorillas. *Oryx* 19: 8–14.

Barns, T.A. 1922. *The Wonderland of the Eastern Congo*. London: Putnam.

Baumgärtel, M.W. 1976. *Up Among the Mountain Gorillas*. New York: Hawthorn Books.

Beebe, B.F. 1969. *African Apes*. New York: McKay.

Benchley, B.J. 1942. *My Friends, the Apes*. Boston: Little Brown.

Bourne, G.H., and Cohen, M. 1975. *The Gentle Giants: The Gorilla Story*. New York: Putnam.

Bradley, M.H. 1922. *On the Gorilla Trail*. New York: Appleton.

Burrows, G. 1898. *The Land of the Pygmies*. London: Pearson.

Burton, R.F. 1976. *Two trips to Gorilla Land and the Cataracts of the Congo*. London: Low.

Cahill, T. and Nichols, M. 1981. Gorilla Tactics. *GEO*. Vol. 3, pp. 101-116.

Cameron, V.L. 1877. *Across Africa*. New York: Harper.

Carr, A. 1964. *Ulendo: Travels of a Naturalist In and Out of Africa*. New York: Knopf.

Du Chaillu, P. 1861. *Explorations and Adventures in Equatorial Africa*. New York: Harper.

Dixon, A. 1981. *The Natural History of Gorillas*. London: Weidenfeld and Nicolson.

Fossey, D. 1983. *Gorillas in the Mist*. Boston: Houghton Mifflin.

Fossey, D. 1970. Making friends with mountain gorillas. *National Geographic*. 137:48–7.

Fossey, D. 1971. More years with mountain gorillas. *National Geographic*. 140:574–85.

Fossey, D. 1981. The imperiled mountain gorilla. *National Geographic*. 159:501–23.

Gatti, A. 1932. *The King of the Gorillas*. New York: Doubleday, Doran.

Goodall, Alan. 1979. *The Wandering Gorillas*. London: Collins Publishers.

Geddes, H. 1955. *Gorilla*. London: Melrose.

Goodall, A.G., and Groves, C.P. 1977. The conservation of eastern gorillas. In *Primate Conservation*, ed. Prince Rainier and G.H. Bourne, 599–637. New York: Academic Press.

Gregory, W. and Raven, H. 1937. *In Quest of Gorillas*. Massachusetts: Darwin Press.

Gregory, W.K. 1950. *The Anatomy of the Gorilla*. New York: Columbia University Press.

Groves. 1970. *Gorillas*. New York: Arco Publishing Co.

Harcourt, A.H. & Fossey, D. The Virunga gorillas: decline of an "island" population. *Afr. J. Ecol.* 19:83–97.

Harcourt, A.H. 1981. Can Uganda's gorilla survive? A survey of the Bwindi Forest Reserve. *Biol. Conserv.* 19:269–82.

Harcourt, A.H. 1980. Gorilla–eaters of Gabon. *Oryx*. 15:248–51.

Johnson, M. 1931. *Congorilla*. New York: Brewer.

Kevles, B.A. 1980. *Thinking Gorillas*, ed. A. Troy. New York: Dutton.

Maple, T., and Hoff, M. 1982. *Gorilla behavior*. New York: Van Nostrand Reinhold.

Merfield, F.G., and Miller, H. 1956. *Gorilla Hunter*. New York: Farrar, Straus & Cudahy.

Raven, H.C. 1931. Gorilla: the greatest of all apes. *Natural History*. 31(3):231–42.

Schaller, G.B. 1961. *The Year of the Gorilla*. Chicago: University of Chicago Press.

Schaller, G.B. 1963. *The Mountain Gorilla: Ecology and Behavior*. Chicago: The University of Chicago Press.

Stewart, K.J. 1977. The birth of a wild mountain gorilla (Gorilla gorilla beringei). *Primates* 18:965–76.

Spinage, C.A. 1970. The ecology and problems of the Volcanoes National Park, Rwanda. *Biol. Conserv.* 26:341–366.

Weber, A.W. & Vedder, A. 1983. Population dynamics of the Virunga gorillas: 1959–1978. *Biol. Conserv.* 26:341–366.

Weber, A.W. 1979. Conservation of the Virunga gorilla. *Wildlife News*. 14(2):7–9.

———. 1981. *Conservation of the Virunga Gorillas: a socio-economic perspective on habitat and wildlife preservation in Rwanda*. MS thesis. U. Wisconsin, Madison.

———. 1987. Socioecologic factors in the conservation of afromontane forest reserves. pp. 205–229 in *Primate Conservation in the Tropical Rainforest* (C.W. Marsh and R.A. Mittermeier, eds.). New York: Alan R. Liss, Inc.

William, Prince of Sweden. 1921. *Among Pygmies and Gorillas*. New York: Dutton.

AFTERWORD

In December 1979, I flew over the Virunga Volcanoes in a vintage DC-3. The grizzled pilot pointed down and said simply, "That's where the American woman lives alone with gorillas." This jolted my memory of the many *National Geographic* stories about Dian Fossey's mountain gorilla studies. I was just beginning my career as a globe-trotting photojournalist on my first journey to Africa (and one of my first anywhere outside of Alabama). The world was jumping at me at an unabsorbable pace.

It took me over a year to convince *Geo's* editors to let me do a "gorilla story." By this time I had read everything available and knew that the mountain gorillas' situation was dire. The massacre of "Digit" and "Uncle Bert" had brought the story to international attention. I came to Rwanda with writer Tim Cahill to do a story about the desperate fight to save a species. What I saw was hopeful but frightening. Rwanda was poor, there were people everywhere and not a wild piece of land in sight. In the middle of this, the Mountain Gorilla Project was setting up a tourism program, anti-poaching was going strong, and an education program was underway.

I stayed two months, walking many miles along Rwanda's lava roads to reach points of entry to the tiny forest that is the only habitat of the mountain gorilla. My increasing knowledge of the world told me that this was the place on earth that best illustrates man's destruction of nature and himself, and where the process could actually be reversed with a real and concentrated effort.

Aperture's publisher, Michael Hoffman, shared my enthusiasm that this book had to be done. George Schaller generously agreed to contribute his voice of true wisdom and authority to this project, lending it a substance that photographs find difficult to claim for themselves.

It is my hope that this book, with its words and pictures, will clarify the gorilla story and serve as potent evidence of the rewards and hazards of the future.

MICHAEL NICHOLS

The following worldwide organizations are active in conservation efforts on behalf of the mountain gorilla. Contributions to assist their important work can be sent to them at the addresses below:

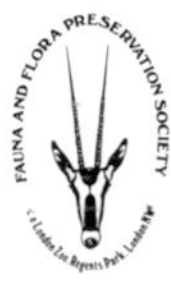

Flora and Fauna Preservation Society
% Zoological Society, London
Regents Park
London NW1 4RY
Attn: John A. Burton, President

AN ORIGINATOR OF THE MOUNTAIN GORILLA PROJECT CONSORTIUM. THEY CONTINUE TO SUPPORT THE EDUCATION PROGRAM IN RWANDA

The Mountain Gorilla Project
African Wildlife Foundation
1717 Massachusetts Avenue
Suite 602
Washington, D.C. 20036

ADMINISTRATES AND FUNDS THE MOUNTAIN GORILLA PROJECT IN RWANDA

Frankfurt Zoological Society
Alfred-Brehm-Platz 16
D-6000 Frankfurt/Main
West Germany

FUNDS THE CONSERVATION OF GORILLAS AND CHIMPANZEES IN EASTERN ZAIRE PROJECT

Wildlife Conservation International
The New York Zoological Society
185th Street and Southern Boulevard
Bronx, New York 10460

SUPPORTS RESEARCH ON THE GORILLAS AND TRAINING OF RWANDAN ECOLOGISTS

Morris Animal Foundation and The Digit Fund
45 Inverness Drive East
Englewood, CO 80112

OPERATES THE VIRUNGA VETERINARY CENTER AND NOW ADMINISTERS THE DIGIT FUND AND THE KARISOKE RESEARCH CENTER

World Wildlife Fund
1255 23rd Street NW
Washington, D.C. 20037
Attn: William A. Reilly, President

FUNDS EDUCATION PROGRAMS IN ZAIRE AND RWANDA AND FUNDS THOMAS BUTYNSKI'S IMPENETRABLE FOREST GORILLA PROJECT IN UGANDA

ACKNOWLEDGMENTS

Thanks to all of those who have given so much of themselves in the effort to save the mountain gorilla, and who have helped me in telling this story, notably:

Mark Condiotti, who came to Africa ten years ago and who has never left, living alone there at great personal sacrifice, his only goal the protection of the gorillas.

Rosalind and Conrad Aveling, who, after leaving the Mountain Gorilla Project (MGP) in Rwanda once it was an established success, set up a mirror program in Zaire, which closed a circle around the Virungas and made the mountain gorilla truly protected.

Jean-Pierre von der Becke, former director of the MGP, who gave up a career in business to devote his life to African wildlife.

William Weber, Assistant Director of Conservation for the New York Zoological Society; Amy Vedder, ecologist and former Karisoke researcher who directed the 1986 census in the Virungas; and who together conceived and initiated the tourism and education components of the Mountain Gorilla Project. Their dedication and efforts in pursuing that idea ensured its success. David Watts, whose central role, over eight years, as a research scientist and former Director of Karisoke was crucial to this success story, and Craig Sholley, newly-appointed director of the MGP, whose tireless work in habituating gorillas and anti-poaching were critical in the early years. (Thanks especially to Bill, Amy, and David for their generous vetting of much of the information in this book).

To Juvenal Habyarimana, President of Rwanda, for his efforts toward preservation of the natural heritage of his beautiful country.

The Rwandan government and the Office Rwandais du Tourisme et des Parcs Nationaux (ORTPN), whose tremendous support has made these advances in conservation possible, as well as the field workers, guides, and guards, too numerous to mention, many of whom have risked their lives to protect gorillas.

The Karisoke trackers and crew, including trackers Alphonse Nemeye, Elias Rukera, Fidele Nshogoza, Emmanuel Rwelekana, Kana Munyanganga, Léonard Munyanshoza, and Antoine Banyangandora, Celestin Nkeramugaba, Salatier Kwiha, Leonard Kananira, Faustin Barabwiriza, and, in camp and on patrols, Andreas Vatiri, Jean-Damascene Ndaruhebeye, and Johan Sekaryongo. True heroes, many of them started working with Dian Fossey as boys, and fought and worked to protect gorillas long before it became fashionable.

Diana McMeekin of the African Wildlife Foundation, whose assistance made much of this possible.

And for their important roles in the study and preservation of the gorillas, as well as generous participation in this project: Alexander (Sandy) Harcourt (former director of Karisoke) and Kelly Stewart, now at Cambridge University, scientists whose research contribution to the study of gorillas has been great; Ann Pierce, former researcher at Karisoke; Tom Lawrence, current MGP anti-poaching officer; Dr. Jim Foster, who established the Virunga Veterinary Center; Lorna Anness and Pascale Sicotte, Karisoke researchers; and gorilla guides "Big Nemeye" (Claver Nemeye), Mathias Kanyamasoro, Jean-Babtiste Basenyurwabo, and Léonidas Zimulinda.

Special thanks to my friend and assistant Shabani.

To George B. Schaller, Director of Science at the New York Zoological Society, whose integrity and abilities are legendary.

Lastly to the memory of Dian Fossey, who dedicated herself to the preservation of the gorilla, setting much of this history in motion.

My very deep personal thanks to: Reba, who has lived with me and worked on this project throughout, administering it all, without whom nothing, ever, would be possible; Philip Jones Griffiths and Donna Ferrato for extraordinary support in times of tribulation; all the staff at Magnum Photos, especially Catherine Chermayeff, who handled the project with her usual aplomb, and Ernest Lofblad, Elizabeth Gallin, Françoise Piffard and Bob Dannin, whose energetic efforts provided timely support; Christiane Brusteadt and Alice Rose George for much-needed advice; Betty Binns, the designer of this beautiful, wonderfully efficient production, and her hardworking, talented associate David Skolkin; at Aperture, Michael Hoffman for his belief in the importance of this book, Steve Baron, whose expertise pulled off a miracle, and Susan Duca, whose brilliant coordination of many details was invaluable, and Nan Richardson, whose creative energy kept us on track and who looked after myriad details under intense pressure.

Thanks to Michael Matson of Canon Cameras, whose generous donation of equipment and support was most appreciated. The photographs in this book were taken with Canon T90 cameras, which have proved their dependability during difficult conditions of rain, humidity, and abuse from Pablo, our "cover gorilla."

To the New Lab of San Francisco, which donated the cost of processing to the project.

To Mike Gurley of Kodak, who donated film for the third trip.

And to Sabena Airlines, especially Mrs. Helen Kahn, who contributed air fare to Rwanda on the latest trip there.

M. N.